D0198222

Good Bird!
A Guide to Solving Behavioral Problems in Companion Parrots!
By
Barbara Heidenreich

Copyright © 2004 by Barbara Heidenreich

Avian Publications
6380 Monroe St. NE
Minneapolis MN 55432

Bruce Burchett Publisher/Owner
www.avianpublications.com
bruce@avianpublications.com
Phone & fax 763-571-8902

All rights reserved. No part of this work covered by copyrights may be reproduced or used in any form or by any means, graphic, electronic or mechanical including photocopying, recording, taping or information storage or retrieval systems, without the prior written consent of the publisher.

National Library of Canada Cataloguing in Publication

Heidenreich, Barbara
Good Bird : a guide to solving behavioral problems in companion birds / Barbara Heidenreich.

ISBN 1-895270-27-8
1. Parrots -- Behavior. 2. Parrots -- Training.
I. Title.
SF473.P3H44 2004 636.6'865 C2003-907325-4

Production and Design
Silvio Mattacchione and Co. / Peter A. Graziano Limited
1251 Scugog Line 8, RR#1.
Port Perry, ON, Canada L9L 1B2
Telephone: 905.985.3555
Fax: 905.985.4005

www.silvio-co.com

Contents
A B C D E F
Hi

Parrots can be some of the most rewarding pets with which to share your life. On the other hand they can also be one of the most frustrating creatures to have in your house! In most cases, parrots exhibit some relatively natural behavior in our homes. Some of these behaviors endear them to us, such as mimicking sounds or preening our hair. However, some other behaviors are absolutely no fun for us at all. These behaviors include screaming and biting, among others.

As a companion parrot owner of 21 years and a professional bird trainer since 1990, I have had the opportunity to meet many parrots and companion parrot owners and witness a great deal of parrot behavior. Over and over again companion parrot owners seek advice for the same, seemingly out of control behavioral problems. The good news is that there are solutions to these problems. The solutions are based on training using positive reinforcement (associating treats or rewards with desired behavior).

Training is really a form of teaching. The subject learns when the trainer positively reinforces (or rewards) each small step the subject takes towards the desired behavior. Using the principles described in training, it becomes easy to teach birds, as well as people, what is the desired response. The beauty of training using positive reinforcement is that it works! In addition, it allows a companion parrot owner to have the best relationship possible with his or her bird.

Scarlet Macaw

In this book I have described some typical problem behavior scenarios experienced by companion parrot owners. I have provided an explanation as to why the behavior may be occurring and offered a step-by-step explanation of how to address the problem behavior using training techniques based on positive reinforcement. If applied correctly, these strategies work. After picking a treat to use as positive reinforcement for your bird and learning about reading bird behavior, feel free to jump ahead to the chapter that most pertains to the specific behavioral problem in which you have an interest.

Over the years, I have worked with a number of birds. However, like most companion parrots owners, I am most fond and familiar with my own blue fronted Amazon parrot. He was an unwanted bird, placed in my home for temporary care that has since turned into 21 years. In the beginning he screamed loudly, was known to bite and was not all that interested in interacting with people. Today he rarely screams. Instead he whistles when he is happy or when he wants something. He has not had a reason to bite in years and will spend hours preening my hair. He has given me many years of joy that have resulted from utilizing the methods described in this book. My goal is for all companion parrot owners to have the opportunity to experience a quality relationship with their birds. It can be a truly rewarding experience.

My Blue Fronted Amazon Parrot, Tarah.

Picking a treat to use as positive reinforcement

In order to offer positive reinforcement or a reward to your bird, it is important to find something your bird really likes. Some birds like head scratches and attention as a reward. However, most birds will respond well to a food treat as a reward.

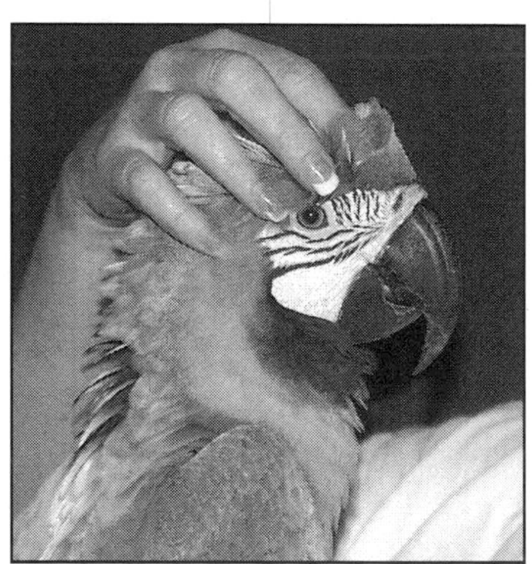

Sometimes a scratch on the head can be used as positive reinforcement.

An easy way to identify food treats for your own bird is to offer your bird his normal diet and observe what food item he eats first. This is typically your bird's favorite food item. This food item can then be removed from the diet and saved for training purposes only. By doing this, your bird will still get the food item, but only for doing desired behavior. Many parrots prefer sunflower seeds, safflower seeds, grapes, banana, corn, peanuts, and other nuts. The seeds are easy to use due to their compact size. If larger food items are chosen, it is recommended those items are either cut or broken into smaller pieces. This allows more opportunities to give your bird a treat for desired behavior before he gets full and no longer has an interest in the treat. Seeds, corn, peas and other small items are usually a good size. However grapes, banana, peanuts and other nuts should be cut or broken into smaller pieces. Avoid using items that are unhealthy for your parrot. High fat items, such as nuts, should be broken into small pieces and used in moderation. I typically use a low oil sunflower seed (grey or sometimes called white stripe-usually available in feed stores), fruits and vegetables and parrot pellets.

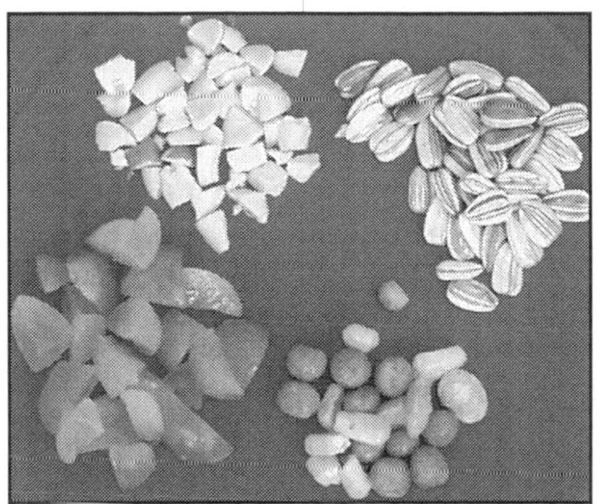

Clockwise from top left: Pieces of peanut, grey stripe sunflower seeds, mixed vegetables, pieces of grape.

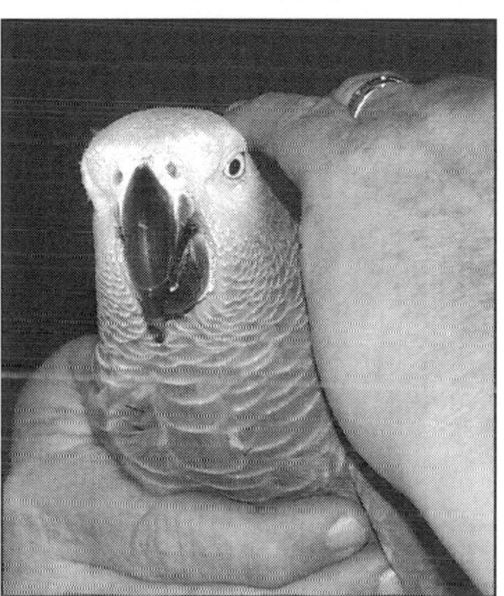

This African Grey Parrot shows body language that indicates he does not enjoy being touched in this manner at the moment.

In order for a treat to be positively reinforcing your bird has to want the treat. If your bird refuses the treat or drops it when offered, try a different treat or try to conduct your session at another time when your bird may have an interest in what is offered. It is especially important when head scratches, toys or attention are used as rewards that your bird

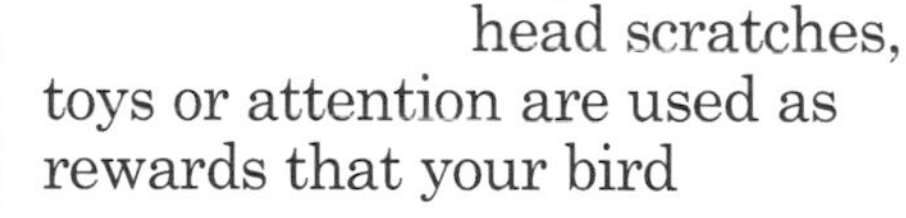

responds in a way that indicates those items are pleasurable to your bird. Trying to scratch your bird's head when he doesn't want to be scratched can hinder progress. The following section describes some basic body language birds can exhibit and what it means. This can help let you know when training is on the right track.

This Umbrella Cockatoo is very relaxed. Note the fluffed feathers under his beak.

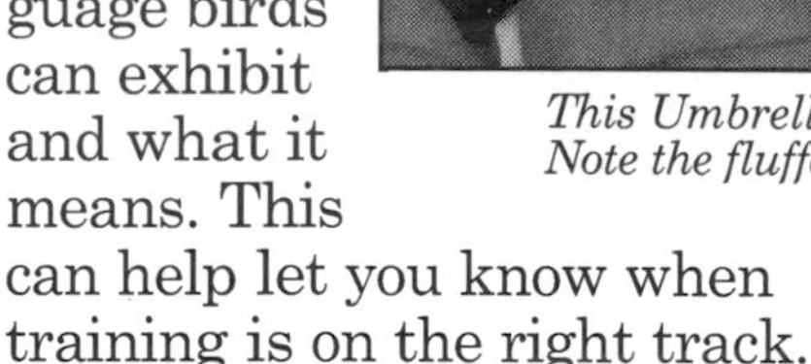

Reading and Interpreting Bird Body Language

The better a companion parrot owner is able to read and interpret bird body language and respond appropriately, the better success he or she will have in solving behavior problems. When training your bird with positive reinforcement, it is important to be sensitive to your bird's reactions. It will be necessary to observe your bird's behavior and respond with the appropriate training choice. The goal is to avoid doing anything your bird perceives as unpleasant. If your bird shows behavior that is indicative of fear and/or aggression, you will need to adjust your actions to avoid this. Again the goal is for all interactions with the companion parrot owner to

This African Grey Parrot is relatively relaxed. His feathers are slightly fluffed. His pupils are normal in size.

be perceived as pleasurable by the bird. This may mean that you may need to stop what you are doing when your bird is uncomfortable with a situation, and try something the bird will accept. To do this, a companion parrot owner needs to know how to read and interpret the body language the parrot is exhibiting. Most companion parrot owners know their birds very well. The following information can confirm what experienced companion parrot owners already know and help those owners who may be new to parrot behavior learn what their parrots are expressing.

Comfort: A relaxed bird may have the feathers on its head slightly fluffed. The bird may also preen, stretch and gently play with a toy or object. A bird that is vocalizing (singing, talking, chirping) is usually comfortable as well. (Note: this is different from screaming) The pupils of the eye appear to be a normal size. If the bird is very relaxed it may hold one foot tucked up close to its body. A bird may also rouse. This is the term used to describe when a bird puffs all of its feathers out then shakes all of its feathers. It takes just a few seconds. A comfortable bird may also shake its tail feathers on occasion and/or flick its wings on its back. Parrots also can sometimes be seen scratching their own heads in a manner that looks like they are giving themselves a head massage.

This Blue Fronted Amazon Parrot is very relaxed. His feathers are very fluffed all over his head and his eye is closing slightly. His head is also low around his shoulders.

Curiosity/investigation: Parrots often use their beaks to explore their world. A young bird especially may be seen putting its beak on many different things. Often a young bird will gradually bite down harder on the object as it investigates it. This can be quite painful if that object is a human finger. If the bird is not fearful, its feathers may be slightly fluffed on its head as it moves in to investigate. A bird may also pick up an object with its feet and hold it while examining it with its beak. An older bird may be more hesitant to examine a new object or person. In this situation the bird may lean back slightly. The feathers may be slicked down and the mouth open slightly. The bird may try to keep its body away from the object while it stretches its neck towards the object to investigate it with its beak. As its fear resides, the bird may move its entire body in closer and relax its feathers.

This young female Eclectus Parrot is investigating the handlers thumb. Note the relaxed body posture and lack of eye pinning.

This Yellow Collared Macaw is showing territorial aggression while sitting in a crate. His feathers are very fluffed (especially on the back of his neck and shoulders) and he is displacing some aggression by biting at the perch.

Aggression: Aggression can occur due to many different reasons. These reasons can include fear aggression, territorial aggression, personal aggression and more. Fear aggression may occur when the bird is afraid and has no escape route.

Note the aggressive body language on this Blue Fronted Amazon. His tail is fanned, wings are held up and away from his body, and his feathers are fluffed at the nape of his neck.

Its only option is to fight back. A good example of this is when a bird is in a cage and is chased by a hand trying to have the bird step up. If the bird is running away from the hand, but is finally cornered in the cage, the bird may bite because it has no option of escape left. Territorial aggression often happens around the bird's primary space. This space could include its cage, a perch, objects (food bowls), another bird, a person or even a room. Personal aggression occurs when a bird consistently displays aggression towards a specific individual. This can be seen when a parrot bonds strongly to one person in the household, and shows aggression towards other members of the household. Typically aggressive behaviors can involve any or all of the following. The bird's feathers may be puffed up on the head and shoulders. The wings may be held out slightly away from the body. At the

This African Grey Parrot looks ready to bite! Again the feathers are fluffed from the nape of the neck down. The wings are held out from the body slightly. This parrot also has his head down slightly, perhaps in preparation to bite.

same time the bird's mouth may be open wider than usual as if preparing to bite. If a bird is very agitated the tail may be fanned out and the pupils will dilate and retract to small dots. This action is commonly known as pinning or eye pinning. Some birds may hiss and/or sway or walk back and forth. Birds without many facial feathers, such as certain macaw species, may become flushed with red on the white exposed skin near the beak and eyes. If a bird intends to bite, often the feathers will slick back on the head and the bird will quickly lunge its head forward towards its bite destination. Some birds may run or fly towards whomever or whatever is the target of the aggression.

Both this Yellow Naped Amazon Parrot and Severe Macaw are displaying eye pinning. Much more of the iris is seen when a parrot pins its eyes. Eye pinning often is displayed when a parrot is excited and/or aggressive.

Excitement: As with other behavior, excitement can have many levels. Excitement can also sometimes lead to aggression. Therefore it is important to keep this is mind when interacting with an excited bird. Signs of an excited bird may include eye pinning, tail fanning, loud vocalizations and/or screaming. Cockatoos will often raise their crests and stand up tall. Birds may be quite active and moving more quickly than usual. Some birds will crouch down and move their wings in short rapid movements against their body.

Fear/nervousness: This is very important body language for the companion parrot owner to know when trying to teach a bird new things using positive reinforcement. It will be a goal to avoid fear and nervousness in training sessions to ensure the sessions are productive. A bird that is afraid will often have its feathers slicked back tight to its body and its eyes will be darting, looking for an escape path. At first the bird may stand tall with its head pulled back a bit. This may change to the bird quickly crouching down and springing up a bit as if it where preparing to launch itself into flight. If it can, the bird will move away from what is making it nervous but try to keep an eye on it while it leaves. If unable to escape it will lean back away from what it fears. It may have

The classic Amazon Parrot strut! When a parrot walks around with his tail feathers spread out, wings out and eyes pinning, he is usually very excited. Keep in mind this also looks very similar to aggression and can sometimes evolve into aggression.

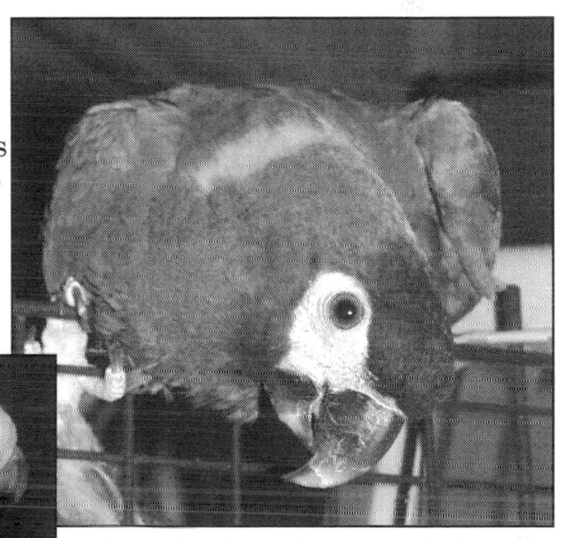

This Yellow Collared Macaw and Umbrella Cockatoo are displaying body language that indicates they are very excited and very anxious. They both are crouched down and wings are quivering. The cockatoo is also screaming. This body language may also be described as begging or desperately wanting something.

its mouth slightly open. The bird could bite if it cannot escape and what it fears continues to approach.

Sick: Birds mask symptoms of illness. This is probably based on instinctual behavior. In the wild a sick bird may be an easy target for a predator. Masking symptoms may help it survive. Typically when birds are not well, they will not behave as they normally would in a situation. This can be very subtle. For example, perhaps your bird usually is quite active in the morning. A sick bird may continue to sleep. Catching a change in behavior early can save a bird's life. Other symptoms may include fluffed up feathers, labored breathing, any noises made while breathing, eyes partially closed, sleeping more than usual, less active than usual and more. Always see a qualified avian veterinarian if any health problems are suspected.

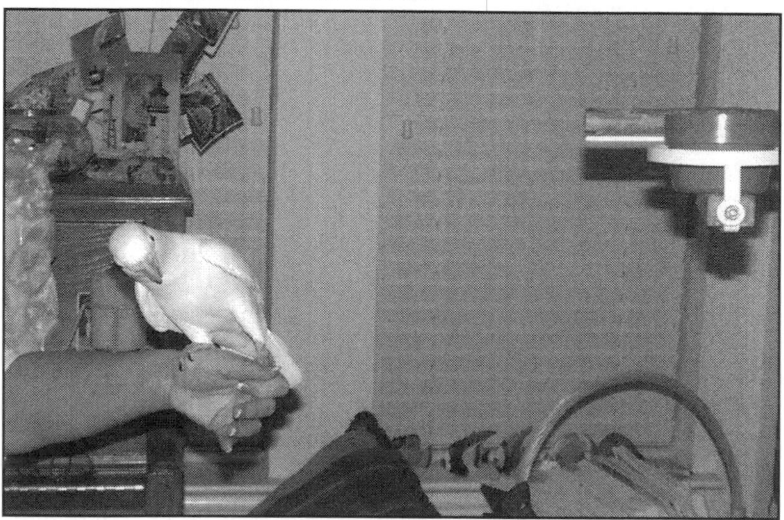

A new perch is very scary to this Goffin's Cockatoo. He is leaning as far away as he can from the perch.

Hot: When a bird is hot, it does not sweat like people do. Instead it will open its mouth and breathe in and out of its mouth to cool off. The bird's feathers will be slicked down. The bird may also hold its wings away from its body. A very warm bird may even hang its head a bit and close its eyes while breathing heavily. Birds can overheat. Take care not to leave birds in situations in which they cannot escape excessive heat. For example, do not leave birds unattended in cars, or place cages by windows in a manner that does not allow the bird to escape direct sunlight throughout the day.

Tired: Often when a bird is getting ready to settle in for the night, the bird will tend to stay in one spot. The bird will fluff its feathers and tuck its head over its shoulder onto its wing. This is often preceded by beak grinding, small head shakes and an occasional small flicking motion made with the wings. The eyelids may look heavy and the bird may blink its eyes often and slowly.

Slightly fluffed feathers, one foot pulled up into the body, slightly closed eyes and a big yawn say this Blue Crowned Conure is ready for a nap.

In general, it is best to work with a bird that is comfortable and relaxed, although curiosity and low levels of excitement can also be helpful to addressing some behavioral issues. Learning to identify the behaviors that indicate fear and aggression will be important to your success. You will want to avoid actions that cause fear or aggression in your bird. Birds that are tired, hot or sick are not in a suitable physical state for training. Make sure their physical needs are met before addressing behavioral problems.

Now that you have your rewards identified and know some bird body language, you are ready to make progress! Look over the following chapters to find the behavioral problem you wish to address.

Parrots scream for a number of different reasons. The key to helping develop a strategy for addressing screaming behaviors is to try to determine why the parrot is screaming in the first place. Different strategies can help different scenarios. The following are a few commonly seen scenarios and strategies that can be applied in those situations. Keep in mind that your bird will probably scream, possibly even more at first, as he learns what you are trying to teach him as you apply these strategies. This is because, up until now, screaming has produced a response from you that your bird wants. Therefore, your bird will continue to try it as an option until he realizes it no longer produces the desired result.

Scenario 1:

You leave the room and your bird starts screaming.

Why does this happen?

Your bird is giving a contact call. He wants to know where you are. What will give him relief is a call back from you or to have you come back in the room. This relates to a parrot's natural strategies for survival in the wild. Most parrots live in pairs and/or flock situations. It is important to stay in touch with your friends to survive if you are a parrot.

Being alone is a very vulnerable condition for a parrot. Many pairs of eyes can be on the look out for predators.

What can you do?

1. Ignore the behavior you do not like (screaming).
2. Reward behavior you do like (silence, or an acceptable sound such as a word or whistle).

Detailed explanation:

1. Ignore the behavior you do not like.

You leave the room and your bird screams. Your bird hopes/expects the scream will get a response from you. If he does not get a response, eventually he will learn screaming does not work. The difficult part about this is that it is very important that you do not do anything that let's your bird think he got a response from you. This may mean freezing in your tracks. If your bird hears you turn on the water, take a step, shut a door, etc. he may take this as enough indication that you are nearby and continue screaming.

Also, when you first implement this strategy, your bird may try to scream excessively in an anxious attempt to get a response from you. However, if you remain patient and give it enough time ignoring the screaming will teach your bird that screaming will not get a response from you.

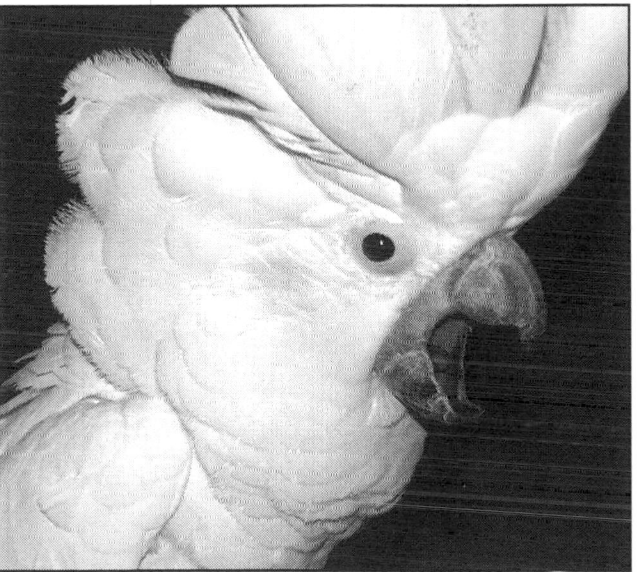

2. Reward behavior you do like.

Because your bird has an instinctive drive to want to

know where you (his parrot buddy) are, it is important to satisfy that need another way. You have decided screaming will not get a response from you. So what will? That is up to you and your parrot to decide. While you are trying to ignore your parrot screaming, he may offer another sound. If he does, and you like that sound, respond to it immediately. You can repeat the sound back, or you can say "good", or you can give your bird visual contact with you, whatever let's your bird know you are there.

You can also reward silence. If your bird stops screaming for a few seconds, respond to that. Then gradually hold out on responding until more time has passed. You will probably have very short time periods (seconds) at first, but can gradually increase those periods over time.

Remember your bird will not get all this right the first time. In the past, screaming has resulted in a response from you. Now the rules have been changed. This is not his fault. He has learned what you may have inadvertently taught him in the past. Now he must learn the new rules. It will take consistent ignoring of screaming and rewarding of desired behavior to solve the problem. However, with consistent practices, a bird can learn not to scream for a contact call easily, within a short period of time.

Scenario 2:

Your bird appears to be frightened of an object and starts screaming anytime he sees it.

Why does this happen?

Being prey items, parrots have an instinctive drive to be wary of new things they don't understand.

What can you do?

1. Desensitize your bird to the new object.
2. Associate good things (rewards) with the presence of the object.

Detailed explanation:

1. Desensitize your bird to the new object.

Desensitizing your bird to the object means allowing your bird to get used to the object. The easiest way to do this is to place the object close enough that your bird can see it, but also far enough away that your bird is not too frightened. Your bird

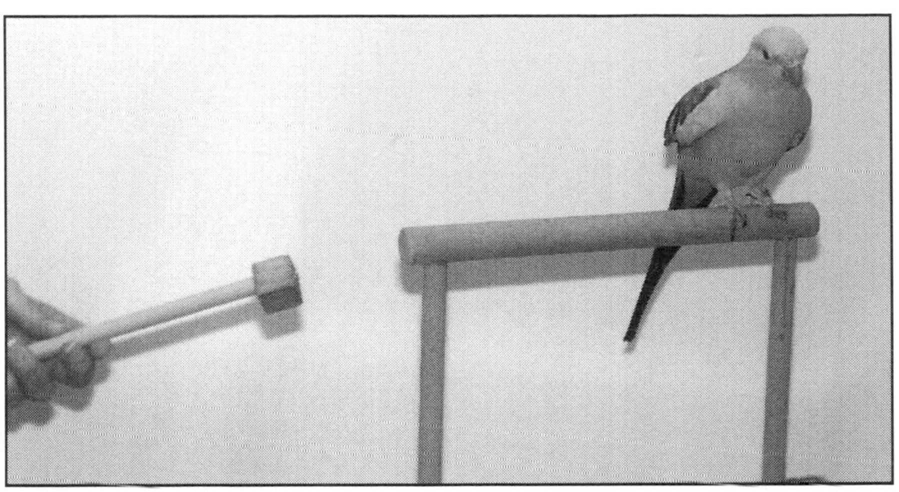

At first, this Rock Pebbler named Kukabird is leaning slightly away from the new object. Kukabird looks at it from the corner of her eye. She looks like she might even fly away if the object were to move closer.

Treats are offered as incentive for Kukabird to move closer to the new object.

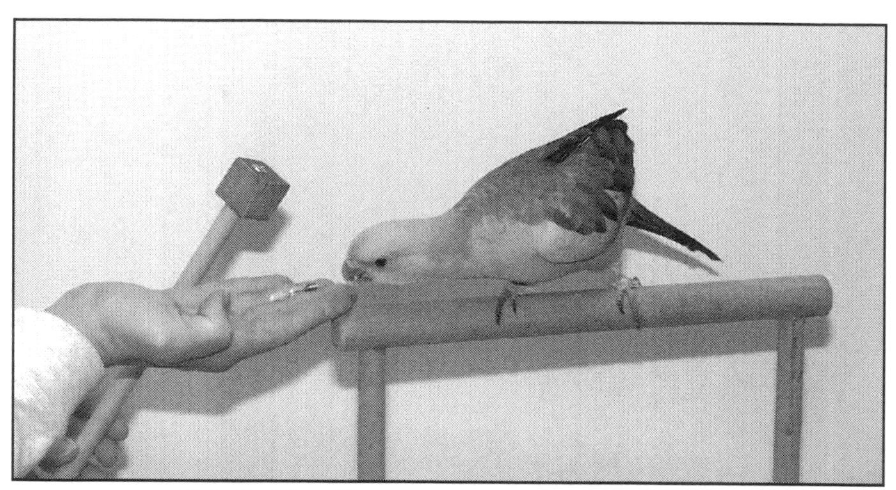

Kukabird is willing to come closer, but not too close.
Notice how she leans over to get the treats.

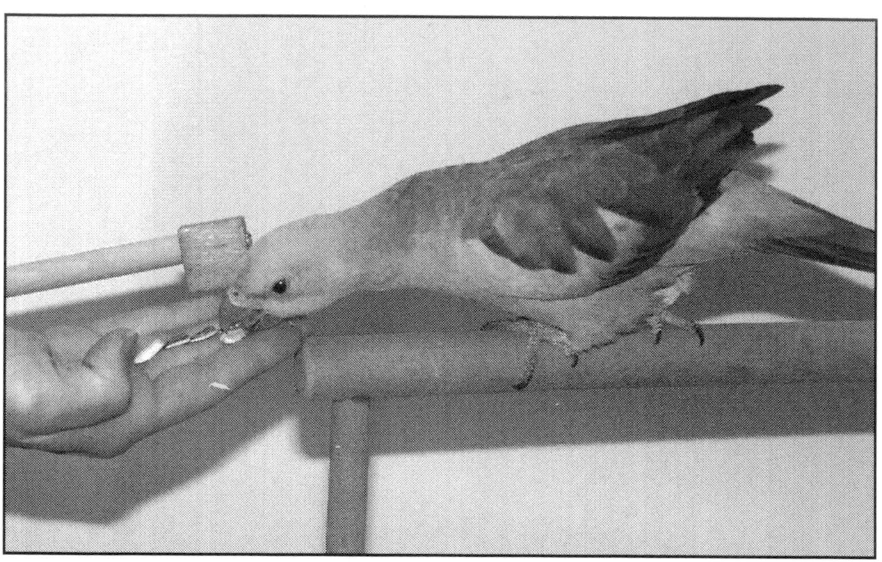

As Kukabird becomes more comfortable with the new object because of the positive reinforcement associated with it, she steps closer.

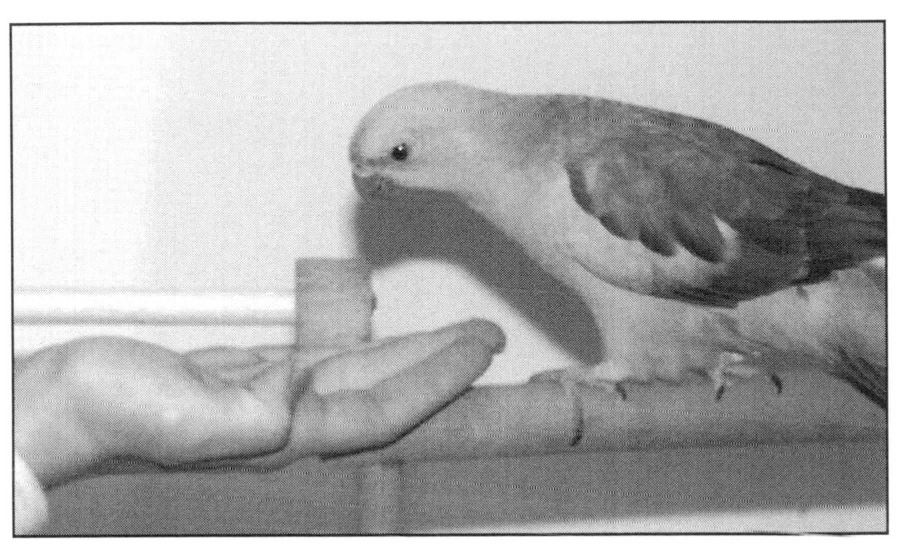

No treats are offered and Kukabird examines the new object without fear.

A treat is placed on the object and Kukabird eats the treat off of the object.

may scream at first, give him a few minutes and see if he will adjust. If not, you may have to move the object farther away. Your bird should be comfortable enough to eat, drink, preen, etc. He may keep one eye on the object, but should still behave relatively normally. Leave the object there to let your bird get used to it. Each day, move the object a little closer, as long as the bird shows calm body language.

2. Associate good things (rewards) with the presence of the object.

As you move the object closer, give your bird a favorite treat. This may take two people. One to move the object, the other to offer the treat at the same time the object is being moved. Pay very close attention to your bird's body language. If your bird shows fear, take smaller steps towards bringing the object closer. The goal is to change your birds' perception of the object as unpleasant and scary to one that is positively reinforcing. The treat being associated with interaction with the object helps to create that change in attitude. By gradually moving the object closer and associating rewards with the object, your bird can become accustomed to the previously frightening object.

Scenario 3:

You are eating dinner and your bird screams, hoping he might get a bite of your food.

Why does this happen?

This bird is screaming because he probably has had at least one occasion when he screamed and received a bite of food. Rather than screaming due to an instinctual drive, the bird has learned screaming gets the desired result of a food treat.

What can you do?

1. Ignore the behavior you do not like (screaming).
2. Reward behavior you do like (silence, or an acceptable behavior such as saying a word or going to a perch).
3. Cue your bird to do the behavior you like and then offer the reward.

Detailed explanation:

1. Ignore the behavior you do not like.

As in previous scenarios, the same principles apply here. When your bird screams for a food treat, do not give him the food treat. In this situation, you don't necessarily need to be quiet or not move. You know your bird wants the food, not to know where you are. So in this case ignoring the behavior means not giving your bird the food treat. As stated before, the screaming may become incessant at first, but will subside when the bird learns screaming will not get the bird his desired result.

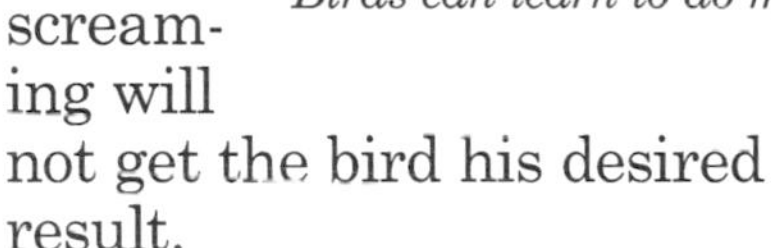

2. Re-ward behavior you do like.

Again, at some point your bird will do something other than scream. It does not necessarily have to involve making any sounds either. Your bird may climb to a perch or scratch his head or hang upside down. If your bird does anything you like, that doesn't involve screaming, give him a treat. Keep in mind whatever he does that you reward will be repeated. You may have a bird that hangs upside down all day long hoping for a treat!

3. Cue your bird to do the behavior you like and then offer the reward.

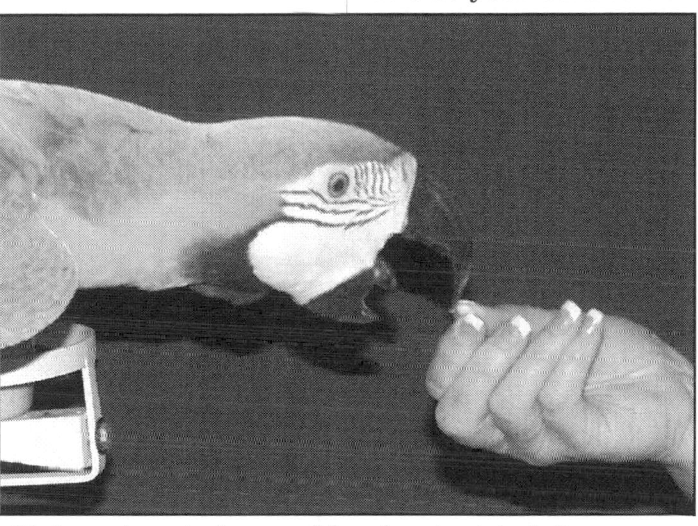

Birds can learn to do many things for a treat, including scream.

There is another strategy that can help move past a bird hanging upside down all day long. That is to put the behavior on cue. Once your bird starts offering the behavior that gets him the treat, start associating a cue with the behavior. You can say a word or give a hand cue while your bird is offering the behavior. If your bird accidentally does the behavior right after you give the cue, give your bird a BIG reward (food treat). This is called a jackpot or magnitude reinforcement. The next step is to try to only give rewards when your bird does the behavior after the cue is given. At this point you stop giving a reward when your bird offers the behavior without the cue. Pretty soon your bird will perform the behavior when you give the cue. Keep in mind you need to ignore (not reward) the behavior if he offers it without being cued first at this point or it will not be clear to him that the behavior must be performed after the cue in order to get the reward. Now when you have dinner, you can cue your bird to perform his behavior on cue and give him his treat for that instead of screaming.

A hand cue tells this Citron Cockatoo to raise his crest and he will get a treat.

Scenario 4:

It is early morning and everyone is asleep, except your bird, who is screaming his head off.

Why does this happen?

In the wild it is typical for parrots to be rather raucous in the morning and the evening. These calls can be about bringing the flock together, defining territory and/or an expression of well-being. Therefore, it is probably a good healthy sign if your bird is noisy in the morning, although not much fun for you.

What can you do?

1. Use a mild form of punishment, not associated with you.

Detailed explanation:

1. Use a mild form of punishment, not associated with you.

In this situation you want your bird to learn that screaming creates an unpleasant circumstance. This is one occasion in which mild punishment may be considered as an option. However, it is essential that the punishment not be associated with you. Usually things that companion parrot owners have been told to do in the past aren't successful (spray bird with water, cover cage, etc.). This is because those things usually involve the companion parrot owner directly and occur well after the scream started. In addition, to some birds, these things are great fun. A more successful approach would be when your bird screams, immediately your bird would experience a mild deterrent that appears to have been caused by your birds own screaming, not by you entering the room.

One family I know had a very loud Umbrella Cockatoo that lived in a renovated garage. The garage was absolutely gorgeous. It was brightly light. It had central heat and air conditioning. It was also elaborately decorated with plants and parrot toys and jungle gyms. And people spent a great deal of time in the area. The garage door was opened on days when the weather was nice to allow natural sunlight and fresh air to circulate. However, the door was opened by pushing a remote control. There-fore, when the door opened it startled the bird, unless someone was standing by his cage to help prepare the bird for the surprise of the door opening. On the other hand, this became a powerful tool to help discourage screaming. Whenever the bird started screaming inappropriately, all the owner had to do was push the remote control garage door opener. Soon the bird learned screaming would cause the garage door to move and therefore discontinued his screaming.

Biting is often taken for granted as an expected part of living with a companion parrot. However, this does not have to be true. Parrots bite for a reason. Determining why a parrot chooses to bite can help solve the problem. Parrots may bite when they have no options left. This can occur when they are being forced to do something they do not want to do. Parrots can also learn to bite for desired response. For example, your parrot may learn that if he is on your arm and would rather be back in his cage, a simple bite may motivate you to put him away. Fortunately, there are positive ways to eliminate biting. The following pages describe some typical scenarios in which a parrot may bite and how to deal with those situations.

Scenario 1:

Your bird is in his cage. You reach in to have him step on your hand and instead he bites your hand.

Why does this happen?

In the simplest terms, your bird does not like what just happened. He may not want to step up, he may have felt that your hand came in too quickly, he may not like the hand pushing on his body, etc. There are many possible "dislikes" about the situation. What is important is that your bird most likely bit, because he had no other option left. If a bird is in a situation in which he cannot leave because he is in a cage or he has his flight feathers clipped, and he does not like what is happening, he has no other choice but to bite to let you know he does not like the situation.

In addition, your bird may have also learned to bite to avoid being picked up. For example, if on one or more occasions your bird bit you when you were trying to pick him up and that resulted in you giving up; he may have learned biting will stop your action of trying to pick him up.

Trying to force a bird to step up can result in a bite.

What can you do?

1. Do not try to make your bird do anything he doesn't want to do.
2. Make stepping on the hand something your bird wants to do using positive reinforce ment.

3. Give your bird a limited amount of time to step up. This will increase the likelihood he will step up when asked. Also, he will step up quicker in the future.

Detailed explanation:

1.Do not try to make your bird do anything he doesn't want to do.

This is a very important concept to learn and apply. Anyone can have a great relationship with a bird, if he or she is careful to respect his or her bird's likes and dislikes. Biting does not have to happen if this simple concept is applied. Before a bird tries to bite he will exhibit other body language. By learning to recognize body language a bird shows when he is uncomfortable, you can learn to stop yourself from moving forward and possibly getting bit. A bird that is thinking about biting may exhibit the following body language. Feathers on the head may be puffed up, mouth may be slightly open. The bird may lean back away from the person. The bird may even lunge with his beak at the person. All these behaviors are the bird's way of saying "no". The bite is the ultimate "no" when all other attempts to tell you he was unhappy with the situation failed. If you see any behavior in which your bird is trying to tell you "no", respect that by not doing whatever it is you are doing that is causing that behavior. There is no need to "dominate" your bird or worry that he is "getting away with not doing the behavior". Birds dislike, are indifferent or like what you are doing. Your goal is to create a situation in which your bird will like what you are doing for 3 reasons. One reason is to avoid ever being bit. Another reason is to create the most positive relationship possible you can have with your bird. And finally both you and your bird will be much happier.

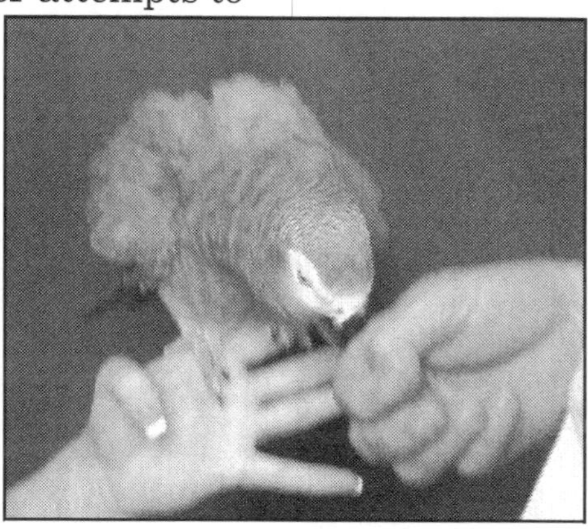

It is easy to see this African Grey Parrot is ready to bite. His body language is saying he does not want to step up on the other hand. Rather than force the bird and possibly get bit, find a way to make it fun for the bird to step up.

2. Use positive reinforcement to make stepping up onto the hand something your bird wants to do.

Now that your bird is not feeling pressured to do something he does not want to do, it is time to find a way to help him want to do the behavior. This means training your bird to step on the hand for positive reinforcement or rewards. The final goal is for your bird to look forward to stepping up on the hand. To create this, it is important to break the act of stepping up on the hand down into smaller steps and reward each step as your bird learns.

Throughout the entire process it is important that your bird is comfortable and relaxed. If your bird is not comfortable and relaxed, you may need to take a step back in your training strategy. The bird's comfort level lets you know how

A hand is rested on the perch next to a Blue Crowned Conure named Dinero.

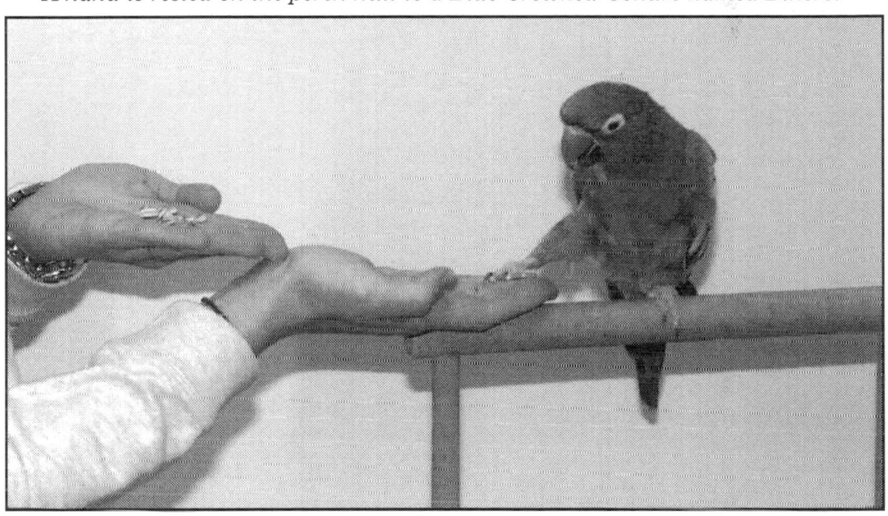

Treats are offered from a hand placed behind the hand resting on the perch. Dinero places one foot on the hand.

far you can go.

For example, if your bird is comfortable, your first step may be to rest your hand on the perch next him. However, if your bird shows fear or aggression, you may want your hand to be farther away or you may need to place treats in the hand resting on the perch. Also, if the bird is very territorially aggressive around his cage, you can try luring the bird with treats to climb onto a portable T shaped perch. Offer the bird a treat as you gently move the perch away from the bird's cage after he has stepped onto the perch. Moving the perch could frighten your bird.

If your bird is comfortable with your hand on the perch next to him, offer your bird a reward. If you are using a food reward you can try bribing your bird to step over your hand resting on the perch to get the treat (sunflower seeds usually work well for this). If your bird moves toward your hand, give him a reward. The next step is for your bird to touch your hand with a foot. Again you can lure your bird over your hand by showing the treats. If he does this, reward him.

The next step is for your bird to place a foot on your hand, reward, then both feet, reward. He may only stay on your hand for a second at this point, but that is OK. Keep your hand steady on the perch and give your bird his reward. The next time allow your bird to step on the hand and off the hand several times in a row and reward him each time. You can then try this with your hand in front of your bird instead of resting on the perch. Once your bird is comfortable with that, try moving your hand away from the perch after your bird has stepped up. This may make your bird very nervous, so allow the bird to go back to the perch if he wants to. Even moving your hand away from the perch needs to occur in steps your bird will accept.

This whole process can occur in one training session (approximately 20 minutes) if your bird has an interest in the rewards you have to offer. If your bird loses interest, you can pick it up where you left off at another time. Getting to the end result is not as important as having the process be as positive as possible. Once the bird learns to step up for positive reinforcement, he will step up quicker in the future. The whole process may seem long, but the end result is a bird that looks for-

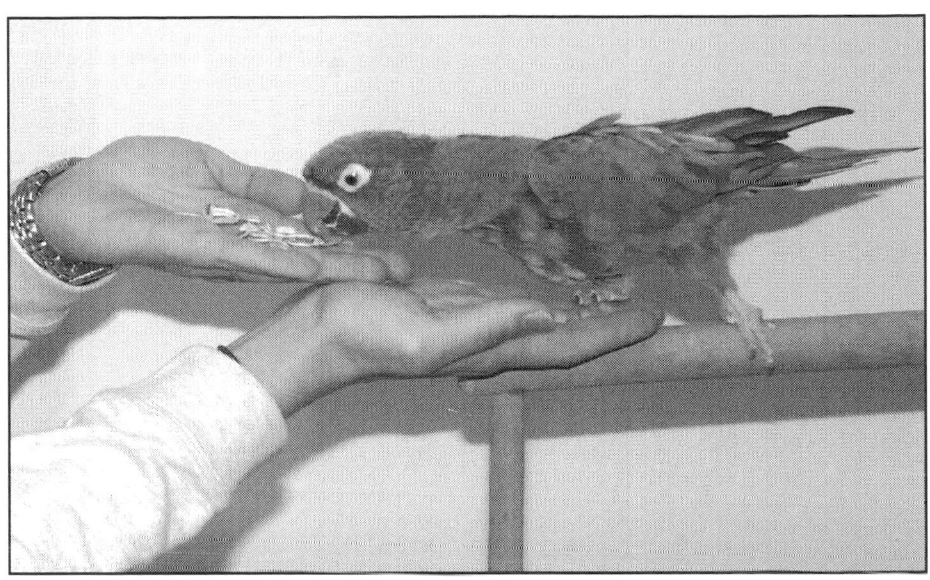

The hand with the food moves in to give him a treat, then returns to its position behind the other hand.

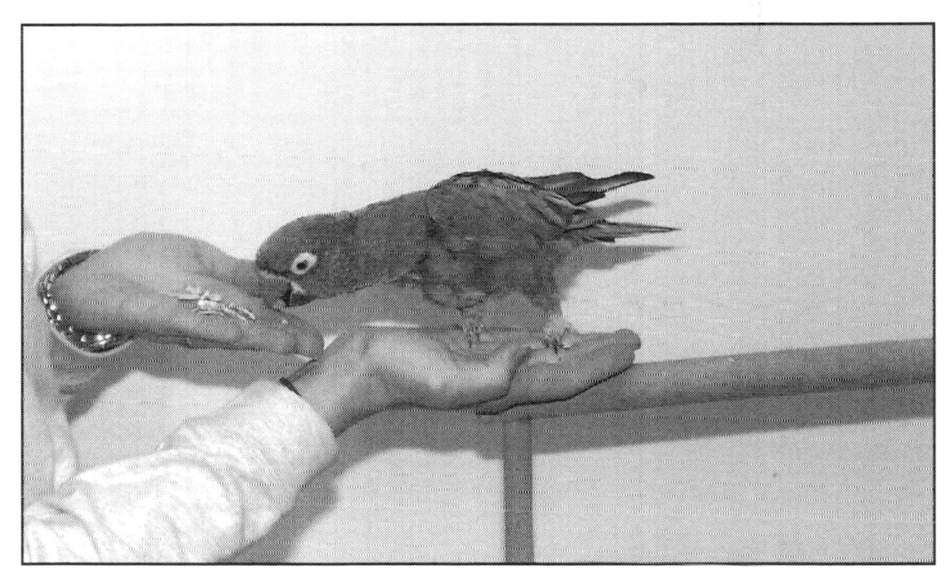

Dinero puts both feet on the hand resting on the perch. He gets several treats (a jack-pot) to let him know that two feet on the hand is very good.

ward to stepping up and sees no need to bite.

3. Give your bird a limited amount of time to step up. This will increase the likelihood he will step up when asked. Also, he will step up quicker in the future.

After your bird has learned how to step up on the hand for positive reinforcement, it is helpful to create quick response to the cue. The cue for your bird to step up is the presentation of a hand in front of his body. If your bird steps up right away (within a few seconds), your bird gets his treat right away as well. That is because your bird performed the behavior in quick response to the presentation of the cue.

If on another occasion you present your hand and your bird gives the body language that he does not want to step up (such as putting his head down, or not lifting a foot to step up, etc.), walk away and take the treats with you. End your training session and try again later when the bird has more interest in the treats. Instead of getting your bird upset and thinking about biting by pushing the situation, you teach him that refusing to step up causes your bird to lose the opportunity to get some tasty treats (of course this only works if your bird wants the treats). After waiting a few minutes, present your hand again. If your bird wants the treat, he will most likely step up quickly this time, rather than miss out on the goodies.

Scenario 2:

Your bird hops down off of his cage, comes after you and tries to bite you.

Why does this happen?

One reason for this behavior may be territoriality. Parrots in the wild have a strong instinct to set up territories. If your bird has formed a pair bond with someone or with another bird in the household, it is possible this bird is doing his best to drive you out of his perceived territory.

What can you do?

(Also see Bonding to One Person)

1.Modify the interactions between your parrot and the person to whom he has bonded.

2.Create positive interactions with the person your bird is trying bite.

Detailed explanation:

(Also see bonding)

1. Modify the interactions between your parrot and the person to whom he has bonded.

The bonded person usually is the one who gets to do all the fun stuff with the companion parrot. This individual can usually pick up your bird, scratch his head, play with him, hand him treats, etc. All these interactions help to strengthen the bond between the person and your parrot. However, this is detrimental to helping your bird to accept another person. During the time you are trying to create a better relationship between your bird and other people, it is best to have the bonded person refrain from such interactions with your bird. The bonded person can clean the cage and put in fresh food and water, however, the less interaction the better.

When cockatoos attack!

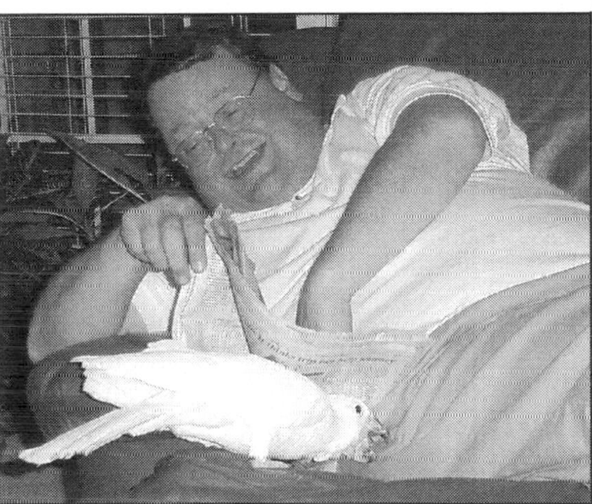

This Umbrella Cockatoo is bonded to the wife in this family and will chase the husband at every opportunity.

2. Create positive interactions with the person your bird is trying to bite.

Because your bird is exhibiting a great deal of territoriality around his cage, it probably would be better to begin to develop a positive relationship away from the cage. If someone in the house is bonded to the parrot, he or she can move your bird to a neutral location. The new person can then begin to work on creating a positive relationship with your bird in the new environment without the bonded person or cage in the vicinity. The positive relationship can be built by having the new person offer the bird treats, toys, head scratches or anything else the parrot enjoys. The idea is to associate things the parrot perceives as "good" with the new person.

After building a positive relationship away from the cage, you can gradually move your bird to a location closer to his cage/territory. Again positive interactions need to occur with the new person in each location. This may be as simple as offering your bird his favorite treats as you gradually get closer to the cage. Remember any behavior that shows your bird is thinking about biting should be respected. Do not push for too much too fast. Also, do not allow the bonded person to be close enough to cause your bird to become aware of that person's presence.

Scenario 3:

Your bird is sitting high above you (perhaps on a curtain rod or on top of his cage) and will not come down. He also tries to bite you when you attempt to pick him up.

Why does this happen?

In the wild there are parrot species that will spend significant time on the ground. However most will also perch as high as they can, especially if they feel threatened or nervous. Perching high gives a parrot an innate sense of comfort that has to do with its strategy for survival in the wild. Parrots are prey items and must be observant of the activities of predators as well as be ready to flee in a moments notice. Perching high offers a good vantage point as well as easily attained lift for flight. Flying off of the ground takes a great deal of energy and can be difficult to do. A parrot is far more vulnerable on the ground than perched high or flying in the air. Therefore perching high can be a very comfortable situation for a parrot. Your bird bites when you try to pick him up because he feels safe and comfortable there. In essence, he is happy where he is and would prefer to remain there rather than step up onto your hand.

This Hyacinth Macaw would much rather play on top of his cage than step up onto the hand.

What can you do?

1. Do not try to make your bird do anything he does not want to do.
2. Make stepping up onto the hand something your bird wants to do using positive reinforcement.
3. Give your bird a limited amount of time to step up. This will increase the likeli hood he will step up when asked. Also, he will step up quicker in the future.
4. Use an alternate positive method to retrieve your bird and train your bird to step up later.

Detailed explanation:

1. Do not try to make your bird do anything he does not want to do.

You may have noticed the same 3 steps are applied here as in Biting Scenario I. This scenario has been included because, although it is the same principles applied, there is a common misconception that because a bird is perched higher, this situation does not follow the same rules of behavior. On the contrary, the exact same theories apply. Again, it is important not to try to force your parrot to step up onto your hand. Your bird will either try to escape from the harassment or if left with no options, he will most likely bite.

2. Make stepping up on the hand something your bird wants to do using positive reinforcement.

If you have already trained your bird to step up onto your hand using positive reinforcement you will have an advantage in this scenario. Obviously your bird is in a different situation, but if he has had enough experiences in which your hand was presented, he stepped up onto it and received a reward; he will be more likely to perform that behavior even when perched above you. In addition, if it is not physically challenging for your bird, and your bird fully understands performing that behavior will earn him a treat, he may climb or fly to get to your presented hand. As in the other scenarios, he also has to want the treat.

On the other hand, if the bird has not been trained to step up in a different situation, this will be a bit more challenging. You can try to train your bird to step up while sitting on the higher perch, but it may be a bit awkward. To begin, it will be easier if you can create a situation in which the bird can step up to your hand rather than down. Parrots seem to be more comfortable stepping up rather than down. Since your goal is success, anything you can do to make it easier for your bird to step up is important. To facilitate this, you may need to stand on a chair or even a ladder. Be careful not to scare your bird if you need to bring a ladder or chair close to him. Move such items slowly and watch your bird's reactions as you bring the item closer.

After you climb the ladder or chair, rest your hand as close to your bird as he will comfortably allow. You can then present treats on the side of your hand opposite to the bird. Essentially you will try to bribe you bird onto your hand with the treats. (See Biting Scenario I for details on training your bird to step up onto your hand using positive reinforcement.) Once your bird is on your hand make sure he gets lots of treats for being on your hand. You may also want to give him lots of treats for going back into his cage. This will help make the cage a positive place to be as well.

Now that your bird is down from the elevated perch you may later want to take some time to practice stepping up for treats to help solidify the behavior of stepping up onto the hand. The more times your bird receives a reward for stepping up, the more likely he will be to perform the behavior in the future.

Using a ladder makes it more comfortable to ask your bird to step up onto the hand. Be sure not to scare your bird with the ladder. Move it in slowly.

3. Give your bird a limited amount of time to step up. This will increase the likelihood he will step up when asked. Also, he will step up quicker in the future.

If your bird knows how to step up onto your hand, but refuses to do so while sitting on the higher perch, close his window of opportunity. In other words, take away his opportunity to get the treat. If your bird understands presenting your hand is his cue to step up onto your hand and his performance of that behavior will earn him a treat, this tool can be useful in this situation. Present your hand and watch his reaction. If he shows no interest, or very little interest, walk away from your bird and leave him there. You can try to present your hand again at a later time. Giving your bird a limited amount of time to do a

A Brazil nut catches the eye of this Hyacinth Macaw.

He decides to climb down on his own to receive the treat.

behavior he knows how to do can increase his motivation to do the behavior. He will learn in order to receive the desired treat and not have you take it away he will need to perform the behavior promptly. Obviously, this strategy works best if you can afford to spend the time presenting the cue, walking away and trying again later.

4. Use an alternate positive method to retrieve the bird and train the bird to step up later.

Sometimes there is not enough time to train your bird to step up onto the hand while he is sitting on the curtain rod, on top of his cage or other high perch. An alternative is to get your bird down as best you can, and work on training the bird to step up later. However, if possible it is best to avoid forcing your bird off of the elevated perch. One method you can try is to move your bird's cage or favorite perch under the bird and allow the bird to climb down to the cage on his own. You can also place favorite treats in your bird's food bowl. At least in this situation, the bird chooses to go to his cage on his own and he will get a treat once he gets there. This helps make inside the cage a more pleasant place to be than on top of it or on another high perch. Again, be sure to work on training the bird to step up later so you will have other options in the future.

Scenario 4

Your parrot was gently mouthing or playing with your finger. However, the pressure of the mouthing began to increase until it became unbearable.

Why does this happen?

Parrots often explore their world with their beaks and mouths. Young parrots in particular may investigate many things with their beaks, including your finger. A hard object (your finger bone) surrounded by soft fleshy stuff could be quite interesting to a curious bird. Up until your bird used too much pressure to explore, he most likely has not had any experience that would discourage that type of exploration. The same applies to a parrot that may be exhibiting play behavior that became a bit too rough for the companion parrot owner.

What can you do?

1. Discourage the excessive pressure by ending the inter action. Give your bird a "time

out".

2. Encourage your bird to per form acceptable behavior and positively reinforce that behavior.

Detailed explanation:

1.Discourage the excessive pressure by ending the interaction. Give your bird a "time out".

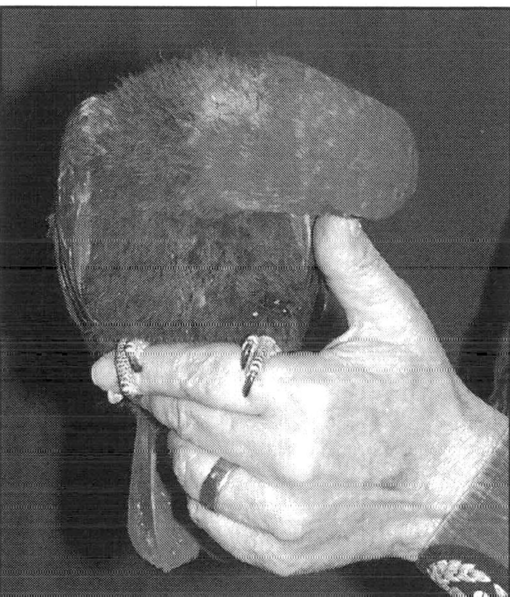

This female Eclectus Parrot is quite fascinated with her handlers thumb.

When your bird is exploring or playing there is motivation to perform these behaviors. The motivation most likely is because your bird enjoys it. If this is true, then ending the opportunity for your bird to explore or play immediately after he has applied too much pressure removes the positive reinforcement he has been receiving from the act of playing/exploring. In other words, if your bird bites too hard during play/exploration you can immediately remove your hands from the play interaction and eventually he will make the connection that the behavior of biting caused him to lose the opportunity to have more fun. Some people call this a "time out". The important thing to remember in this situation is that there is no need to anything more than gently remove your hands. Your bird does not need to be yelled at, have a finger waved in his face, be tapped on the beak, etc. to learn mouthing too hard is not acceptable. From your bird's perspective, he was just playing and exploring.

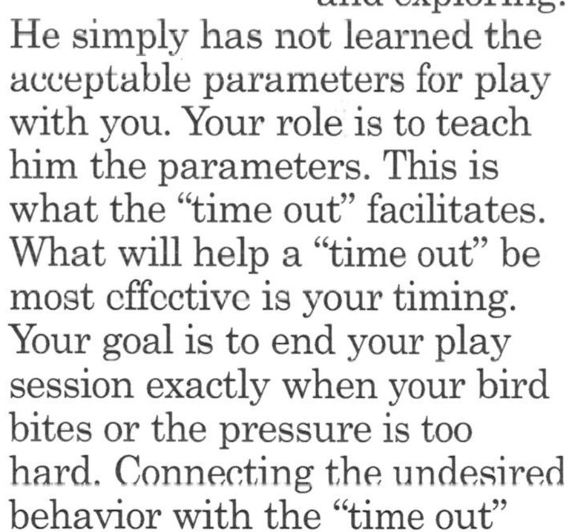

He simply has not learned the acceptable parameters for play with you. Your role is to teach him the parameters. This is what the "time out" facilitates. What will help a "time out" be most effective is your timing. Your goal is to end your play session exactly when your bird bites or the pressure is too hard. Connecting the undesired behavior with the "time out"

helps the bird to learn exactly what was perceived as an unacceptable behavior.

2. Encourage your bird to per form an acceptable behavior and positively reinforce that behavior.

A "time out" does not have to last a long time. A few short seconds may be enough for your bird to understand what behavior lost his opportunity to interact with you. After those seconds have passed, you can pick up your bird again and resume your activity. As long as your bird does not bite too hard and is enjoying the interaction, you are positively reinforcing appropriate play behavior by interacting with him. If the play gets too rough again, you can always give your bird another "time out" and repeat the process.

Another option is to cue your bird to perform a behavior he already knows and positively reinforce your bird for performing that behavior. For example, your bird bites too hard during play and you give him a time out. When the time out is over, cue your bird to wave (or some other behavior he knows how to perform on cue) and either pick him up to play again, scratch his head if he enjoys that or offer a food reward. The purpose of having your bird perform an acceptable behavior is to keep your relationship as positive as possible. It is possible to teach your bird what behavior is unacceptable without having to be harsh, as well as teaching him what behavior is acceptable. In addition, you maintain a relationship with your bird based on positive interactions.

Another way to encourage appropriate play behavior is to redirect your birds focus on acceptable play behavior. You can gently remove your finger from your bird's mouth and offer another acceptable play object to your bird. There is no reason to assume your bird knows what objects are approved for play and what are not. Set you and your bird up for success by choosing a play area that is free of "off limit" toys. Provide plenty of suitable toys for your bird to enjoy as acceptable options. This helps to avoid the temptation to use aversives with your bird that might occur if your bird chose to play with something that is not perceived to be a toy by you.

Scenario 5

Your bird is sitting on your arm and suddenly bites you.

Why does this happen?

Witnessing the act often gives a clearer picture as to why the event occurred. However there are two explanations that often seem to be applicable to this situation. One explanation is redirected aggression. Redirected aggression occurs when a bird is demonstrating aggressive behavior in regards to another person, bird, object, etc. However the bird cannot get to the thing that has him agitated. Instead he aggresses on the closest object. This can be a perch, the cage bars, or your arm, among other things. This is often seen when a parrot is perched on the arm of the person to whom he has bonded, and another person walks into the room. The parrot may exhibit aggressive behavior about the person entering the room, but bites the person to whom he is bonded in an effort to redirect aggression. (Biting can also happen if the person holding the bird is not the person the bird prefers, and the bonded person walks into the room. (See Bonding to One Person)

Another common explanation for biting in this situation is because the parrot has learned to bite to get a desired response from the companion parrot owner. For example, often if a companion parrot owner is bitten by his or her bird, the bird is usually put back into his cage. It is possible that this bird could learn that biting will allow the bird to go back to his cage. If this is what the bird wants, he can quickly learn to bite in order to train the companion parrot owner to put the bird away. This can also work in the other direction as well. If for example, the bird is brought back to his cage and rather than go in, he reaches down to bite. If the companion parrot owner's reaction is to pull the bird back away from the cage, and this is what the bird wants, the bird has learned biting will allow him more time spent with his owner. It only takes one experience in which a bird bit and got a desired response in order for the bird to learn biting can get him what he wants.

What can you do?

1. Pay attention to your bird's body language. Prevent redirected aggression by avoiding situations that increase aggressive behavior.

2. Be aware of what you do when your bird bites.

3. Teach your bird using posi

tive reinforcement to do what you would like him to do.

Detailed explanation:

1. Pay attention to your bird's body language. Prevent redirected aggression by avoiding situations that increase aggressive behavior.

If your bird is biting in an attempt to redirect aggression, try to avoid situations that agitate your bird. For example if you have noticed your bird's behavior often becomes more aggressive when a certain individual is in the room, make sure your bird is not perched on your arm during times when that individual is present. This may mean alerting the household you plan on having your bird out on your arm for awhile and to stay clear. Often parrots will also get excited when there is a lot of activity and commotion. Sometimes loud conversation on the phone or in the household may create a level of excitement that could turn to aggression. Perhaps the phone ringing or knowing your children are about to come home with friends should be cues for you to not have your bird on your arm. Which scenarios might cause redirected aggression will vary with individual birds. To identify what situations to avoid will require you to read and interpret your bird's body language. If your bird begins to show any behaviors that indicate aggression (see Reading and Interpreting Bird Body Language), calmly put your bird away before he gets to the point of biting. Time spent with your bird under more suitable conditions will allow for a more positive interaction for the both of you.

2. Be aware of what you do when your bird bites.

If your bird bites before you could avoid it, take note of what he does and what you do. Does your bird bite and then look towards a place he might like to go, such as his cage or another person? In this case the bird may have learned to bite for a desired response.

In the past, what has been your reaction? If a bird bites you, it not necessary for you to "take the bite" to teach your bird he cannot get away with biting. However, it also not necessary to throw, hit, or yell at your bird either. Because it can be quite painful, the first thing to do is to stop the bite if possible. If your bird is holding on, this may mean grasping the top of your bird's beak and prying it off of you to release the pressure. You may then just put your bird down wherever you can; the couch, the floor, a table and walk away. Let your bird calm down before you try to pick him up again.

If your bird looks as if he is preparing to bite, you can possibly prevent the bite. If you can identify why he might want to bite, you can avoid the situation. For example, if you notice someone coming into the room that may cause your bird to displace aggression, you can ask the person to wait a moment while you put the bird away. If you cannot avoid the situation and your bird is still leaning over to bite, you can redirect your bird onto an acceptable behavior such as targeting. It is difficult for your bird to stay balanced on your arm and bite at the same time. This can potentially distract your bird until you can either get your bird back to his cage or alter the situation to avoid aggressive behavior.

If your bird has given you a quick bite, you may want to consider a "time out". (See Biting Scenario IV). This may mean putting your bird back in his cage or perhaps leaving him perched anywhere but on you. An important factor to keep in mind is that the time out is only effective if he wants to be doing what he was doing prior to the bite. If he bit in hopes of returning to his cage, and he is returned to his cage, you will have taught him to bite for a desired response.

3. Teach your bird using posi tive reinforcement to do what you would like him to do.

What if your bird bites as you approach his cage because he does not want to go back in to his cage? To address this situation, you can teach your bird going back into his cage is fun by using positive reinforcement. This may mean giving him his favorite treat for going into his cage. There are several techniques you can use here.

One technique is to hold the treat in your hand and offer it to your bird as you approach the cage. Allow the bird to hold onto the treat with his beak, while you also continue to hold onto the treat with your hand. Ease the bird through the cage door. Once through the door, let go of the treat so your bird can enjoy it inside his cage. Doing this gives you a positive way to get a reluctant bird through the doorway. You can also try just dropping the treat into a food bowl set in the back of the cage.

This may bribe the bird into the cage. You can also show your bird the treat, as your hand goes to the back of the cage to lure your bird into the cage. Eventually your bird will learn that your hand is a target. If he follows that target he will receive a treat. This is a very useful tool. If your bird learns to view your hand as a target, you can eventually hide the treat in your hand. This will allow you to avoid bribing your bird into the cage.

The reason to avoid bribing after the behavior is initially trained is that your bird may learn to wait to see if you will offer bigger and better treats and therefore not perform the behavior. If you hide the treat in your hand and also change what the treat may be, so that your bird is surprised, you increase the likelihood that your bird will perform the behavior of going into the cage. Offering different types and quantities of treats is called variable reinforcement and it can increase motivation.

Giving this Hahn's Macaw a treat as he went through the doorway helped him look forward to going back into his cage.

It is similar to reaching into a cookie jar that contains different types of cookies. You cannot see what types of cookies are inside. When you reach inside to grab a cookie, you may pull out your favorite kind. The next time perhaps it is not your favorite, but one you still like. However, you are willing to try again, because next time you may pull out your favorite again. As long as you are hungry for cookies, you are still willing to reach into the cookie jar.

These two birds are not interested in stepping up on the hand at the moment.

However, they are willing to go back into their cage for a few treats dropped in their food bowl.

Chapter 4: Bonding to One Person

Scenario 1:

Your parrot only likes one person in the house. You may be the one who cleans the cage, feeds the bird, etc. However, your bird suddenly will do anything for someone else in your household and shows only aggressive behavior towards you.

Why does this happen?

Bonding to one individual is a very natural behavior for a parrot. In the wild, parrots form very strong pair bonds with their mates. Once the bond is formed, each bird will defend his territory vigorously. Only the mate will be allowed into the territory (usually the cage for a companion parrot). If another parrot approaches that bird may be aggressively driven away.

What can you do?

1. Modify the interactions between your parrot and the person to whom he has bonded.

2. Create positive interactions with other members of the household.

Detailed Explanation:

1. Modify the interactions between your parrot and the person to whom he has bonded.

The bonded person usually is the one who gets to do all the fun stuff with the companion parrot. This individual can usually pick up your bird, scratch his head, play with him, hand him treats, etc. All these interactions help to strengthen the bond between the person and your parrot. However, this is detrimental to helping your bird accept another person. During the time you are trying to create a better relationship between your bird and other people, it is best to have the bonded person refrain from such interactions with your bird. The bonded person can clean the cage and put in fresh food and water, however, the less interaction the better.

2. Create positive interactions with other members of the household.

In order to allow another person to have a successful interaction with your parrot, the bonded person must not be around. This will help curb those natural urges your parrot has to drive the competition away. One way to help build a positive relationship is to have the new person drop a treat in your bird's food bowl and then walk away. If this is done often enough, eventually your bird will look forward to the new person coming by with a treat. This person can then try offering the treat to your bird through the cage bars. Eventually the treat can be offered for stepping up on the hand as well as other cooperative behavior. If your parrot is very territorial around his cage towards the new person, the bonded person can bring your bird to a place in which he doesn't exhibit territorial behaviors. Next, have the bonded person leave the room. The new person can then enter the room and try the process of offering treats in the new environment. The new person can also cue your bird to perform simple behaviors. By doing quick repetitions of simple behaviors your bird knows how to do, your parrot will have an activity on which to focus. He will also have many opportunities in which he will receive positive reinforcement from the new person. This can help build a positive relationship with the new person. In the beginning the scale of positive interactions is in favor of the bonded person. The new goal is to bring the scale back in balance by limiting the positive experiences with the bonded person and increasing the positive experiences with the new person. This strategy can help make an unbearable situation livable. However, keep in mind that it may never be possible for the new person to interact at the same level the bonded person can with your bird, and it is unlikely positive interactions can occur while the bonded person is in the room. By keeping track of levels of aggression and bonding behaviors (such as regurgitating to share food), parrot owners can adjust how much positive attention your bird receives from different people in order to maintain a balance of accepted behavior with different members of the household.

I Luv U?
A B C D E F
Hi

Scenario:

Your bird has spent almost his entire life inside of his cage. He is now deathly afraid to come out.

Why does this happen?

When parrots are young they are often very accepting of new things and new environments. Baby parrots spend a great deal of time learning how to survive in the wild from their parents. The baby parrots are exposed to a variety of circumstances and learn the appropriate response from their parents. In a captive situation, a young bird is also more at ease with change. Parrot owners can also use this time frame to expose a young bird to different environments, foods, people, etc. If each of these experiences is positive for the bird, the odds are in favor of the bird adjusting well to new situations, including coming out of its familiar cage. However, a bird that never has had that opportunity, especially when young, may be very fearful of such a new situation.

What can you do?

1. Make coming out of the cage a positive experience for your bird.

Detailed explanation:

1. Make coming out of the cage a positive experience for your bird.

A great way to overcome fear is through positive reinforcement. To help explain how this would work, we will make a goal for your cage bound bird to sit on a perch that is ten feet away from the cage. Again it is important to break the behavior down into very small steps. If your bird doesn't know how to step up on your hand for positive reinforcement, training this behavior first would be very helpful (See Biting Scenario I). If your bird will sit on your hand, you can offer a treat as you slowly try to move your hand out of the cage. If he appears frightened, do not go quite as far quite as quickly.

As you come closer to the door of the cage, your bird will probably have more difficulties. At first going through the door will most likely be the most challenging part of the behavior. One way to facilitate going through the door is to hold a treat in your fingertips and allow your bird to hold onto it with his beak. Do not let go of the treat and bring your bird through the door. As soon as you clear the door, let your bird have the treat. (See Biting Scenario V).

The next step may seem strange, but it is important. Immediately put your bird back in the cage. By doing this, your bird isn't forced to cope with more than he can possibly handle at this point. Repeat the above steps several times. The more times your bird goes through the door and receives positive reinforcement, the less frightening it will become to your bird to go through the door.

After your bird appears comfortable going through the door, you want to add distance away from the cage to the equation. The steps you have taken have been small up until now. If your bird ever shows a fear response, allow your bird the opportunity to return to the cage, and then go through your approximations again. You can gradually add distance to this and eventually get to the new perch 10 ft away from the cage. Remember, your bird may also be afraid of the new perch and you may need to gradually allow your parrot to get used to the new perch by offering treats to your bird when he is near the perch or actually steps up on it.

This cage bound Umbrella Cockatoo, Harrison, was recently found starving in an abandoned home. His new caretakers have given him a good home and have been working with him to overcome his fear of leaving his cage. Harrison loves a good head scratch, but still needs the security of his cage nearby. Notice how he hangs on with one foot.

Once your bird has become accustomed to coming out of his cage you may want to increase time that your bird spends in new environments. Again, approach these changes by taking small steps with your bird and making sure each step is rewarded with positive reinforcement.

Read your bird's body language and avoid behavior that indicates fear (darting looks, slicked back feathers, crouching to spring into flight, etc.) by taking small enough steps and ending your training session on a positive note. Each positive step will take you closer to your goal.

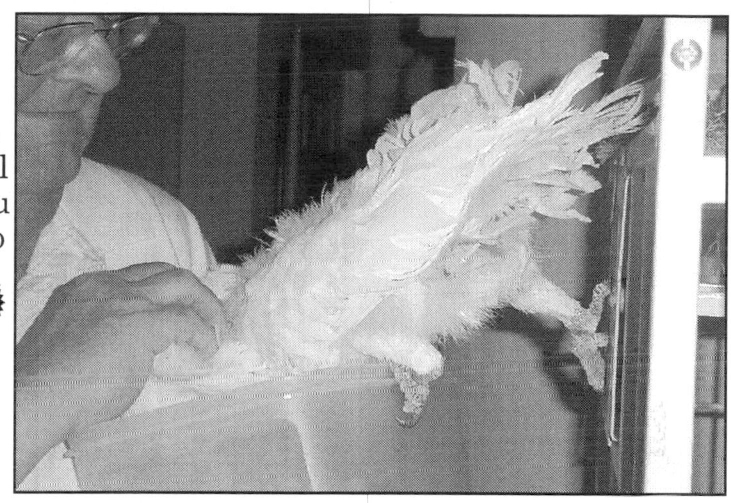

Scenario:

Your bird has taken preening to a whole new level. Rather than taking care of his feathers, your bird destroys feathers, leaving bare patches of skin. Some birds go as far as mutilating skin.

Why does this happen?

It is difficult to ascertain why birds pick their feathers. In some cases there are specific medical causes. Therefore it is always important to have your bird examined by an avian veterinarian. A veterinarian who stays current in the latest information on avian medicine is highly recommended.

If a medical cause has been ruled out, it is time to consider a behavioral cause and a behavioral solution. Sometimes change in environment can cause this destructive behavior. Sometimes other stress inducing situations may be involved. Sometimes no obvious factor initiated the behavior.

Identifying the cause may be helpful, but once the behavior has started, what initiated it may no longer be of importance. Sometimes what initiated the behavior is no longer occurring. However, what keeps the behavior occurring over and over again continues. This is the self-reinforcement the bird experiences when he performs the behavior. For example, many people bite their fingernails or pick at scabs. Although both of these behaviors are self destructive, people will continue to perform those behaviors. This is because in some way, those behaviors offer something positive. That positive thing may be the release of endorphins that make the person feel better on some level. This could also be what a bird experiences when he picks his own feathers. However, there is not conclusive evidence of this at this point in time.

What can you do?

1. Distract your bird from performing destructive behavior.
2. Cue and reward your bird for doing a different non self-destructive behavior.
3. Use preventative measures. Cue your bird for good behavior before he starts to pick.

Detailed Explanation:

1. Distract your bird from performing the destructive behavior.

One way to address feather picking behavior is to get your birds attention off the self-destructive behavior. This may require startling your bird mildly. It is crucial that this startle occurs while your bird is picking feathers. The timing is critical. It is important your bird makes the connection that the startle is only associated with the behavior you do not want. The startle can be something extremely mild like a flashlight being flashed against a nearby wall. It is amazing how birds react to a strange light moving seemingly all by itself. A quick loud noise your bird will take note of may also work. This may involve dropping something, or blowing a whistle quickly, etc. Any action taken to startle your bird should be quick and preferably not associated with you. It only needs to distract your bird. If it doesn't distract your bird immediately, don't use it. In addition, overtime methods used to distract may lose their effectiveness. You may have to constantly change your devices to get your birds attention. Also, be careful not to go overboard and seriously frighten your bird. Pay attention to his reaction.

Sometimes it is not clear why some birds feather pick and some do not. Pictured here is a pair of Cherry Headed Conures.

2. Cue and reward for doing a different non-destructive behavior.

Immediately after your bird's attention has been diverted from the feather picking, cue him for another behavior you like. If your bird performs the behavior, reward him with a treat or head scratch, etc. If your bird does not know a behavior on cue, reward him for doing anything other than picking, even if he just sits there not picking for a few seconds, give him a reward. Remember, just about any non-destructive behavior will do. Your bird may do something as simple as scratch his head or walk a little. Any behavior other than feather picking is acceptable.

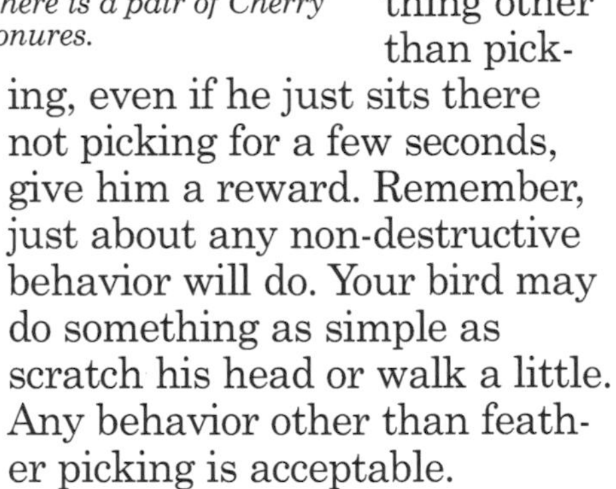

3. Use preventative measures. Cue your bird for good behavior before he starts to pick.

Distracting and cueing for good behavior works after your bird has started picking. However, even more effective would be trying to teach your bird to do something else before he starts picking. Pay close attention to your bird's behavior. If you can notice any pattern of behavior that occurs right before your bird starts to pick, you can cue your bird for a good behavior and offer him a reward just before the picking would have started. Perhaps your bird scratches a lot, or starts shaking his head. The signs may be very subtle. But if you can identify them, you may be able to prevent picking with a positive interaction between you and your bird.

A flashlight directed at the wall, may momentarily take your bird's mind off of chewing his feathers. Be careful not to frighten your bird. Your goal is just to redirect his attention.

Note: Feather picking can be very challenging to correct. The methods described above obviously can work if a person is around. Unfortunately, because feather picking can be self-reinforcing, there are times when it may occur when your bird is by himself. There are many other suggestions that may help, including providing toys, regular baths, diet changes, moving your bird to a new location, etc. It is recommended to consider those options as well.

Here are some final general recommendations to keep in mind:

✐ Take the time to focus on making the process of learning as positive as possible. It may take time to reach the final goal, but in the long run both you and your parrot will benefit.

✐ View your parrot's behavior as a product of what he has learned. Understand he is only doing what has worked for him in the past. If you want to change that behavior, you will need to teach him what behaviors are acceptable. There is no need to label him as a bad bird. Instead try the strategies described in this book.

✐ Learn the difference between allowing a bird to choose to do something and forcing a bird to do something. Sometimes force can be so small, we do not realize we are forcing. For example, force is involved when a person slides his finger under a birds toe and lifts the bird off of a perch. This may be very little force, but it is still force.

Holding the finger completely still and allowing the bird to lift his foot to step onto the finger is allowing the bird to choose. Try to allow a bird to do something instead of forcing.

There are many ways to train or teach parrots to be good companions in our homes. The methods presented in this book are not the only methods. Nor do they cover each and every situation an individual may encounter.

I recommend you learn as much as possible from as many resources as you can. Choose what methods make sense to you and appear to be safe and kind for your bird. I do hope you have found suggestions in this book that will be useful to you and your companion parrot.

Barbara Heidenreich has been a professional in the field of animal behavior since 1990. After receiving her Bachelors of Science degree in Zoology from the University of California at Davis, she began her career in 1990 teaching birds to present natural behaviors in zoos at what was then known as Marine World Africa USA in Vallejo, CA.

She later developed educational programs featuring birds at Brookfield Zoo near Chicago, IL. She then relocated to Disney's Discovery Island in Orlando, FL to continue training birds for free flight educational programs in a zoological setting.

Other experience includes 5 years working for Steve Martin's Natural Encounters, Inc. as Director of Operations for this company she helped produce over 15 different free flight bird presentations. These included the Flights of Wonder Show at Disney's Animal Kingdom and the Birds of the World Show at the State Fair of Texas among others.

Her experience includes consulting on animal behavior and training in zoos around the world, teaching bird training workshops and presenting talks to bird clubs and other organizations. She is now a private animal training consultant working with zoological parks, nature centers and the companion parrot community.

More information on her services can be found at www.GoodBirdInc.com and www.ATandCS.com.

She is the Past President of the International Association of Avian Trainers and Educators (IAATE), as well as the Conservation Committee Chairperson. She is a charter member of the Animal Behavior Management Alliance (ABMA) and long time member of the American Zoo and Aquarium Association (AZA).

Barbara has always had a passion for animals and especially for the unique bond that can be created between birds and people. She hopes everyone can experience that truly rewarding relationship that can happen when animals are trained using positive reinforcement.

Barbara is pictured here with a Burchell's Coucal.

As with any project, there are many people, and in my case animals, to thank. First and foremost I need to thank, Tarah, my Blue Fronted Amazon who has taught me so much and has been an incredible companion for many years. I also owe a great deal to Lea and Adam Czyzweski for their ongoing support and friendship, through good times and bad. A special thanks to Jaye Apperson for her encouragement, support, knowledge, wisdom and editing skills! I would also like to extend my gratitude to Mary Healy, Donna Sue Evans, Rebecca O'Connor and Brad Shaffer, Ph.D. for their continued support over the years. Special thank you to Terry and Sam Debrow for making me laugh until I cried while we shot photographs. And thank you also to Kathy and Bill Ferguson for more photo opportunities as well as providing a number of "parrot models" for this book. A special thank you goes to Jean and Jim Gibson for digital equipment usage, editing services, good advice and general acts of kindness.

I would also like to thank Steve Martin and Susan Friedman, Ph.D. for sharing their knowledge and expanding mine. A special thanks to the International Association of Avian Trainers and Educators, Natural Encounters, Inc. staff members and Parrots and People.

Finally, I would like to thank all the parrots I have had the privilege to know. They all have taught me so much. And to all the companion parrot owners looking for ways to shape behavior, thank you for making a positive relationship with your parrot a priority.

Good Bird Magazine

Published quarterly by Good Bird Inc. Good Bird Magazine is the ultimate resourse for individuals seeking to learn kind and gentle ways to create desired parrot behavior Our primary focus is behavior from an applied behavior analysis approach with an emphasis on positive reinforcement training. However we also cover other topics related to the wellbeing of parrots in our homes and in the wild.
Available at
www.goodbirdinc.com

Don't Shoot the Dog! The New Art of Teaching and Training

By Karen Pryor
(1999 Bantam Publishers)

This book is essential to anyone's collection. It discusses some basic training principles in very simple language. The training principles are then applied to everyday situations to help understand how they can be used with animals and people. It is easy to read and inexpensive as well.
Available through
www.clickertraining.com,

www.amazon.com and in most major bookstores in Psychology and Self Help sections.

Animal Training: Successful Animal Management through Positive Reinforcement

By Ken Ramirez (1999 Shedd Aquarium Publishers)

This book is a collection of articles written about animal training. It is very comprehensive and includes a great deal of information. Articles cover a variety of animal species. It is a large book and a bit more expensive, but worth it. It is available by calling toll free 1-888-732-7883 (1-888-SEA-STUF) or visiting www.sheddnet.org.
This book is also available through www.amazon.com.

Living and Learning with Parrots: The Fundamental Principles of Behavior

Online Class

By Susan G Friedman, Ph.D.

This online class is designed to provide companion parrot owners with a deeper understanding of the processes that influence behavior and associated teaching strategies. The goal is to build and maintain a successful relationship with your feathered companions proactively rather than reactively.
To register, send an email to Susan at behavior101@comcast.net. For more information visit www.parrottalk.com.

The Cambridge Center for Behavioral Studies

This organization helps people view behavior from a scientific perspective. The website offers forums to discuss behavior. Visit www.behavior.org for more information.

Parrots & People

A non profit organization that exists to facilitate long term care for parrots in need of placement via foster care and adoption programs. Potential families receive extensive instruction and support before birds are placed in their care, as well as follow up support. Parrots & People also provides conservation and educational programming for special needs people and retirement facilities. Visit www.parrotsandpeople.org or call toll free 1-866-540-0057.

The Parrot Problem Solver. Addressing Aggressive Behavior.

By Barbara Heidenreich

Learn how to modify your parrot's aggressive behavior. Includes information on parrot behavior in the wild, 10 steps to address aggressive behavior, common myths and misconceptions about parrot behavior, real life case studies and their solutions.
Available at
www.goodbirdinc.com or
www.avianpublications.com

An Introduction to Parrot Behavior and Training DVD

By Barbara Heidenreich

The first in our series of Parrot Behavior and Training DVD's, this DVD covers the basic fundamentals of training with positive reinforcement. We give you detailed instructions on how to shape behavior. Our training subjects are companion parrots learning behaviors for the first time. Be the first to experience these special moments of learning captured on film.
Available at
www.goodbirdinc.com or
www.avianpublications.com

Train Your Parrot for the Veterinary Exam DVD

By Barbara Heidenreich

Make your bird's next visit to the veterinarian stress free! This DVD shows you how to train your parrot to cooperate in his own medical care. Step by step instructions demonstrate how to train each behavior.
Available at
www.goodbirdinc.com or
www.avianpublications.com

Aggression:

Offensive or defensive behavior in response to aversive stimuli in which a bird may exhibit any or all of the following behaviors; feathers may be puffed up on the head and shoulders, wings may be held out slightly away from the body, mouth may be open wider than usual as if preparing to bite, the tail may be fanned out, eye pinning may occur, birds may hiss and/or sway or walk back and forth. Birds without many facial feathers may become flushed with red on the white exposed skin near the beak and eyes. If a bird intends to bite, often the feathers will slick back on the head and the bird will quickly lunge its head forward towards its bite destination. Some birds may run or fly towards whomever or whatever is the target of the aggression.

Aversive:

Something that the subject does not like and will work to avoid.

Beak grinding:

A behavior parrots typically exhibit when relaxed and close to falling asleep. The upper and lower parts of the beak are gently rubbed against each other repeatedly in grinding action. Usually short grinding noises are followed by a small pause, and then repeated.

Biting:

Any action in which the mouth is used to apply pressure to an object or person.

Blood feather:

A feather that is still growing. It has a blood and a nerve supply. Damage to a blood feather either through intentional clipping or accidental contact can cause pain and bleeding to the bird and should be treated by an avian veterinarian.

Bribing:

Showing the subject the positive reinforcement (treat or reward) it will receive prior to and/or during the subject performing the behavior. Liability of using bribing is that the subject may decide the bribe is not sufficient to perform the behavior.

Bridge (bridging stimulus):

A sound or signal that the subject has learned that means the behavior has been performed correctly and positive reinforcement will be presented. It is called a "bridge" because it bridges the gap in time between when the behavior was performed and when the subject will receive the positive reinforcement. An effective bridge should be easily distinguished from other sights and sounds the subject may experience. Some bridges commonly used by animal trainers include the word "good", a whistle, and a clicker.

Clicker training:

Training in which the sound of the clicker is used as the bridge. Some people prefer using a clicker as a bridge because the sound is very distinct and sounds the same every time. It can be a very clear signal to the subject.

Clipped (wing or feather):

A condition in which any or all of the primary and/or secondary flight feathers have been cut in order to prevent a bird from obtaining lift to fly. Once a feather has grown in completely it no longer has a blood or nerve supply. Trimming feathers that are completely grown in does not cause the bird pain. Blood feathers should not be clipped as they can bleed and can be painful for the bird.

Contact call:

A sound a parrot may make when it is trying to locate another parrot or person.

Crest feathers:

Feathers found on the top of a parrot's head. Some species of parrots have longer crest feathers, such as some cockatoos. These feathers may be raised and/or lowered in response to different stimuli in a parrot's environment.

Cue:

A sound or signal that tells the subject what to do.

Desensitize:

The process of allowing the subject to become more comfortable with an object, person and/or situation over time that previously produced a negative response.

Displaced aggression:

The action of performing aggressive behavior on or towards anything other than the stimulus that is causing the aggression.

Dominate:

Overpowering a subject physically or mentally in order to exert control over the subject.

Extinction burst:

Excessive and often powerful performance of a previously positively reinforced behavior in order to gain reinforcement. It occurs just prior to the extinction of the behavior.

Eye Pinning:

Condition in which the pupils of a parrot dilate and restrict to tiny black dots. This displays a great deal of the colorful iris in some parrot species.

Height dominance:

Controversial theory that indicates parrots establish a dominance hierarchy based on the height at which they perch. Parrots that perch higher are said to be dominant. Opponents believe height dominance in parrots does not exist. Parrots perch higher for an opportunity for better visibility, better lift for flight, comfort, etc.

Hiss:

Sound parrots may emanate from the mouth when displaying aggressive behavior. More commonly seen in cockatoo species.

Instinct:

Behavior that does not need to be learned. It is believed to have a genetic component. Also known as hard wired behavior, innate or species specific behavior. Some examples in birds include nest building, predator avoidance behaviors, and courtship.

Jackpot:

Positive reinforcement that is perceived as better in quality and/or quantity by the subject. It is offered when a subject has performed a behavior very well. The trainer uses a jackpot to make a big impression on the subject that what was presented was correct. Also known as magnitude reinforcement.

Limited opportunity:

After the presentation of a cue, the subject is allowed a few seconds to perform the behavior it knows how to do. If the behavior is not performed promptly, the opportunity to gain positive reinforcement is removed. This increases the likelihood the behavior will be performed quickly when the cue is presented.

Magnitude reinforcement:

This is positive reinforcement that is better in quality and/or quantity. It is offered when a subject has performed a behavior very well. The trainer uses a magnitude reinforcement to make a big impression on the subject that what was presented was correct. Also known as a 'jackpot'.

Motivation:

Desire or drive. Reason for behaving.

Negative reinforcement:

A consequence that occurs in conjunction with the performance of a behavior that increases the likelihood that the behavior will be performed again and is also something the subject does not like. The subject will only work to the level necessary to avoid the aversive stimulus.

Neutral location:

Any area that would not be considered part of its territory by the subject.

Operant Conditioning:

A science that focuses on how subjects learn. It involves a stimulus or antecedent, followed by the performance of a behavior followed by a consequence. The consequence determines whether the behavior will be repeated or not. Negative reinforcement, positive reinforcement, punishment are all elements of operant conditioning.

Pair bond:

A situation that naturally occurs between two parrots in the wild, often with a mate. Two parrots will intentionally spend most of their time together. A bonded pair may exhibit any or all of the following behaviors; preen each other, regurgitate for each other, fly in synchronicity, sit side by side, behave aggressively towards other birds of the same species and more. This bond can be shared with a human instead of a bird in captive situations. This is very likely to occur with a hand raised parrot that spends the majority of its time with one person.

Personal aggression:

Aggressive behavior that appears to be directed towards a specific individual under most circumstances. (Note this is different from territorial aggression and aggression due to a bond with another person. Personal aggression can happen away from the territory and also when the person to whom a parrot is bonded is not in the vicinity)

Pin feather:

When a feather is growing it appears to have a "paper" like opaque wrapper around it called a horny sheath. This is normal and is typically preened off by the parrot or its mate or the companion parrot owner. While the sheath is still on the feather, the feather is known as a pin feather.

Positive reinforcement:

A consequence that occurs in conjunction with the performance of a behavior that increases the likelihood that the behavior will be performed again and is also something the subject likes. If the subject wants the positive reinforcement offered, the subject will often work harder to gain the reinforcement.

Preening:

Behavior in which a bird will use its beak to take care of its feathers. This can include grasping the feather with the beak at the base of the feather then pulling the beak over the feather to the tip of the feather. The behavior reconnects feather barbs that have come apart. It can also be done to remove the horny sheath that is found on a feather that is growing or has just finished growing. Preening also spreads oil over the feathers secreted by a gland at the base of the tail. Preening can be done to the bird by itself, to another bird and sometimes to a companion parrot owner.

Rouse:

A behavior usually exhibited by a bird that appears to be relaxed. Typically the bird raises all the feathers on its body than shakes the head, body and tail. The behavior last only a few seconds and appears to be related to making sure feathers are in good condition.

Screaming:

Vocalization by a parrot that is exceptionally loud and somewhat monotone. It can be repeated in short bursts and can be associated with distress and/or fear, but it can also be learned to be performed to get desired results such as attention, food, etc.

Target:

A tool that can be used to help shape behavior. Almost anything can be a target. Typically a subject is trained to go to or follow a target by associating positive reinforcement with the target. Example: A dog may follow its food bowl because it has received positive reinforcement (food) in the bowl.

Teaching:

Method of communicating to change behavior. By applying certain communication techniques, the subject can learn what the desired response is. Teaching can be intentional or unintentional. Synonymous with training.

Territorial aggression:

Aggressive behavior that occurs in conjunction with an area, object or person that is perceived to be part of the bird's territory. Example: cage or food bowls.

Territorial behavior:

Behavior that indicates a particular area, object or person is exclusive to that individual bird. It appears to be an innate or hard wired behavior for parrots to establish an area that is to be used exclusively by that parrot and its mate for survival of the species.

Time out:

Temporarily ending a training session or removing the opportunity for the subject to perform behavior to receive positive reinforcement.

Training:

Method of communicating to change behavior. By applying certain communication techniques, thc subject can learn what the desired response is. Synonymous with teaching. Training presented in this book is a form of applied behavior analysis called operant conditioning.

Variable reinforcement:

Providing different amounts and/or different types of positive reinforcement after each performance of a behavior. Varying the reinforcement increases the subject's motivation to perform the behavior because the subject never knows what it might receive.

Vocalizing:

Sounds a parrot makes when it is relaxed and comfortable that involve using its voice. This can include mimicking words, sounds, mumbling, singing, etc. This does not refer to screaming.

The End!
A B C D E F
Hi

Good Bird!

THE COMPLETE PET BIRD OWNER'S HANDBOOK

(Revised Edition)

by Dr. Gary A. Gallerstein

Exclusively from Avian Publications, and written by a veterinarian for bird owners, this book covers selection, nutrition, behavior, home physicals, emergency medical care, preventative medicine, achieving optimum health, and understanding what your bird needs from you for a happy, healthy life. Includes an index of health and illness signs, a bibliography of additional resources, and superb photos and illustrations. **Hardcover. 432 pp.**

"WOW!
This book is a gem. Packed with information and good insight, it needs to be a part of every pet bird owner's library. Dr. Gallerstein has done it again!"

Brian Speer, DVM; Diplomate, ABVP, ECAMS Co-author, 'The Large Macaws' and 'Birds for Dummies'

Visit us at www.avianpublications.com

Made in the USA
Charleston, SC
05 December 2012